A Brief Guide to

MOLD,

MOISTURE,

AND

YOUR HOME

EPA 402-K-02-003
(Reprinted 09/2012)

This Guide provides
information and guidance
for homeowners and
renters on how to clean
up residential mold
problems and how to
prevent mold growth.

U.S. Environmental Protection Agency
Office of Air and Radiation
Indoor Environments Division
1200 Pennsylvania Avenue, N. W.
Mailcode: 6609J
Washington, DC 20460
www.epa.gov/iaq

A Brief Guide to Mold, Moisture, and Your Home

Contents

MOLD **BASICS**

- **The key to mold control is moisture control.**

- **If mold is a problem in your home, you should clean up the mold promptly *and* fix the water problem.**

- **It is important to dry water-damaged areas and items within 24-48 hours to prevent mold growth.**

Why is mold growing in my home? Molds are part of the natural environment. Outdoors, molds play a part in nature by breaking down dead organic matter such as fallen leaves and dead trees, but indoors, mold growth should be avoided. Molds reproduce by means of tiny spores; the spores are invisible to the naked eye and float through outdoor and indoor air. Mold may begin growing indoors when mold spores land on surfaces that are wet. There are many types of mold, and none of them will grow without water or moisture.

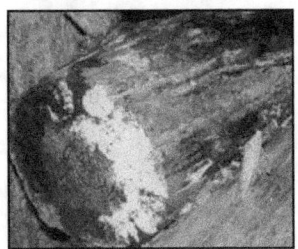

Mold growing outdoors on firewood. Molds come in many colors; both white and black molds are shown here.

Can mold cause health problems? Molds are usually not a problem indoors, unless mold spores land on a wet or damp spot and begin growing. Molds have the potential to cause health problems. Molds produce allergens (substances that can cause allergic reactions), irritants, and in some cases, potentially toxic substances (mycotoxins).

Inhaling or touching mold or mold spores may cause allergic reactions in sensitive individuals. Allergic responses include hay fever-type symptoms, such as sneezing, runny nose, red eyes, and skin rash (dermatitis). Allergic reactions to mold are common. They can be immediate or delayed. Molds can also cause asthma attacks in people with asthma who are allergic to mold. In addition, mold exposure can irritate the eyes, skin, nose, throat, and lungs of both mold-allergic and non-allergic people. Symptoms other than the allergic and irritant types are not commonly reported as a result of inhaling mold.

Research on mold and health effects is ongoing. This brochure provides a brief overview; it does not describe all potential health effects related to mold exposure. For more detailed information consult a health professional. You may also wish to consult your state or local health department.

How do I get rid of mold? It is impossible to get rid of all mold and mold spores indoors; some mold spores will be found floating through the air and in house dust. The mold spores will not grow if moisture is not present. Indoor mold growth can and should be prevented or controlled by controlling moisture indoors. If there is mold growth in your home, you must clean up the mold **and** fix the water problem. If you clean up the mold, but don't fix the water problem, then, most likely, the mold problem will come back.

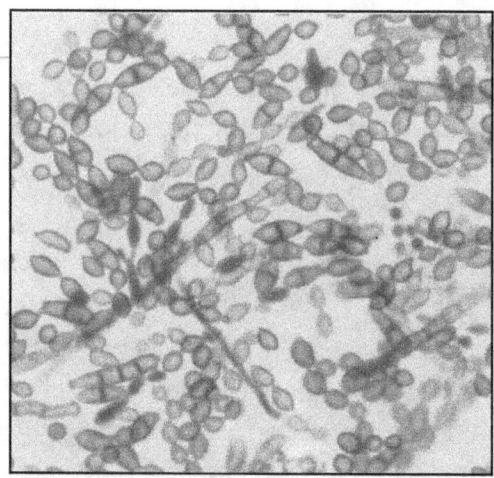

Magnified mold spores.

Molds can gradually destroy the things they grow on. You can prevent damage to your home and furnishings, save money, and avoid potential health problems by controlling moisture and eliminating mold growth.

MOLD CLEANUP

Leaky window – mold is beginning to rot the wooden frame and windowsill.

If you already have a mold problem –

ACT QUICKLY.

Mold damages what it grows on. The longer it grows, the more damage it can cause.

Who should do the cleanup? Who should do the cleanup depends on a number of factors. One consideration is the size of the mold problem. If the moldy area is less than about 10 square feet (less than roughly a 3 ft. by 3 ft. patch), in most cases, you can handle the job yourself, following the guidelines below. However:

■ If there has been a lot of water damage, and/or mold growth covers more than 10 square feet, consult the U.S. Environmental Protection Agency (EPA) guide: *Mold Remediation in Schools and Commercial Buildings*. Although focused on schools and commercial buildings, this document is applicable to other building types. It is available on the Internet at: www.epa.gov/mold.

■ If you choose to hire a contractor (or other professional service provider) to do the cleanup, make sure the contractor has experience cleaning up mold. Check references and ask the contractor to follow the recommendations in EPA's *Mold Remediation in Schools and Commercial Buildings*, the guidelines of the American Conference of Governmental Industrial Hygenists (ACGIH), or other guidelines from professional or government organizations.

■ If you suspect that the heating/ventilation/air conditioning (HVAC) system may be contaminated with mold (it is part of an identified moisture problem, for instance, or there is mold near the intake to the system), consult EPA's guide *Should You Have the Air Ducts in Your Home Cleaned?* before taking further action. Do not run the HVAC system if you know or suspect that it is contaminated with mold - it could spread mold throughout the building. Visit www.epa.gov/iaq/pubs to download a copy of the EPA guide.

■ If the water and/or mold damage was caused by sewage or other contaminated water, then call in a professional who has experience cleaning and fixing buildings damaged by contaminated water.

■ If you have health concerns, consult a health professional before starting cleanup.

BATHROOM TIP Places that are often or always damp can be hard to maintain completely free of mold. If there's some mold in the shower or elsewhere in the bathroom that seems to reappear, increasing the ventilation (running a fan or opening a window) and cleaning more frequently will usually prevent mold from recurring, or at least keep the mold to a minimum.

Tips and techniques The tips and techniques presented in this section will help you clean up your mold problem. Professional cleaners or remediators may use methods not covered in this publication. Please note that mold may cause staining and cosmetic damage. It may not be possible to clean an item so that its original appearance is restored.

■ Fix plumbing leaks and other water problems as soon as possible. Dry all items completely.

■ Scrub mold off hard surfaces with detergent and water, and dry completely.

Mold growing on the underside of a plastic lawnchair in an area where rainwater drips through and deposits organic material.

Mold growing on a piece of ceiling tile.

▣ Absorbent or porous materials, such as ceiling tiles and carpet, may have to be thrown away if they become moldy. Mold can grow on or fill in the empty spaces and crevices of porous materials, so the mold may be difficult or impossible to remove completely.

▣ Avoid exposing yourself or others to mold (see discussions: **What to Wear When Cleaning Moldy Areas** and **Hidden Mold**.)

▣ Do not paint or caulk moldy surfaces. Clean up the mold and dry the surfaces before painting. Paint applied over moldy surfaces is likely to peel.

▣ If you are unsure about how to clean an item, or if the item is expensive or of sentimental value, you may wish to consult a specialist. Specialists in furniture repair, restoration, painting, art restoration and conservation, carpet and rug cleaning, water damage, and fire or water restoration are commonly listed in phone books. Be sure to ask for and check references. Look for specialists who are affiliated with professional organizations.

WHAT TO WEAR WHEN CLEANING MOLDY AREAS

Mold growing on a suitcase stored in a humid basement.

It is important
to take
precautions to
LIMIT YOUR EXPOSURE
to mold and
mold spores.

■ **Avoid breathing in mold or mold spores**. In order to limit your exposure to airborne mold, you may want to wear an N-95 respirator, available at many hardware stores and from companies that advertise on the Internet. (They cost about $12 to $25.) Some N-95 respirators resemble a paper dust mask with a nozzle on the front, others are made primarily of plastic or rubber and have removable cartridges that trap most of the mold spores from entering. In order to be effective, the respirator or mask must fit properly, so carefully follow the instructions supplied with the respirator. Please note that the Occupational Safety and Health Administration (OSHA) requires that respirators fit properly (fit testing) when used in an occupational setting; consult OSHA for more information (800-321-OSHA or osha.gov/).

8

- **Wear gloves**. Long gloves that extend to the middle of the forearm are recommended. When working with water and a mild detergent, ordinary household rubber gloves may be used. If you are using a disinfectant, a biocide such as chlorine bleach, or a strong cleaning solution, you should select gloves made from natural rubber, neoprene, nitrile, polyurethane, or PVC (see **Cleanup and Biocides**). Avoid touching mold or moldy items with your bare hands.

Cleaning while wearing N-95 respirator, gloves, and goggles.

- **Wear goggles**. Goggles that do not have ventilation holes are recommended. Avoid getting mold or mold spores in your eyes.

How do I know when the remediation or cleanup is finished? You must have completely fixed the water or moisture problem before the cleanup or remediation can be considered finished.

- You should have completed mold removal. Visible mold and moldy odors should not be present. Please note that mold may cause staining and cosmetic damage.

- You should have revisited the site(s) shortly after cleanup and it should show no signs of water damage or mold growth.

- People should have been able to occupy or re-occupy the area without health complaints or physical symptoms.

- Ultimately, this is a judgment call; there is no easy answer.

MOISTURE AND MOLD PREVENTION AND CONTROL TIPS

MOISTURE
Control is the Key to Mold Control

Mold growing on the surface of a unit ventilator.

■ When water leaks or spills occur indoors - **ACT QUICKLY**.
If wet or damp materials or areas are dried 24-48 hours after a leak or spill happens, in most cases mold will not grow.

■ Clean and repair roof gutters regularly.

■ Make sure the ground slopes away from the building foundation, so that water does not enter or collect around the foundation.

■ Keep air conditioning drip pans clean and the drain lines unobstructed and flowing properly.

Condensation on the inside of a window-pane.

■ Keep indoor humidity low. If possible, keep indoor humidity below 60 percent (ideally between 30 and 50 percent) relative humidity. Relative humidity can be measured with a moisture or humidity meter, a small, inexpensive ($10-$50) instrument available at many hardware stores.

■ If you see condensation or moisture collecting on windows, walls or pipes - ACT QUICKLY to dry the wet surface and reduce the moisture/water source. Condensation can be a sign of high humidity.

Actions that will help to reduce humidity:

♦ Vent appliances that produce moisture, such as clothes dryers, stoves, and kerosene heaters to the outside where possible. (Combustion appliances such as stoves and kerosene heaters produce water vapor and will increase the humidity unless vented to the outside.)

♦ Use air conditioners and/or de-humidifiers when needed.

♦ Run the bathroom fan or open the window when showering. Use exhaust fans or open windows whenever cooking, running the dishwasher or dishwashing, etc.

Actions that will help prevent condensation:

◊ Reduce the humidity (see preceeding page).

◊ Increase ventilation or air movement by opening doors and/or windows, when practical. Use fans as needed.

◊ Cover cold surfaces, such as cold water pipes, with insulation.

◊ Increase air temperature.

Mold growing on a wooden headboard in a room with high humidity.

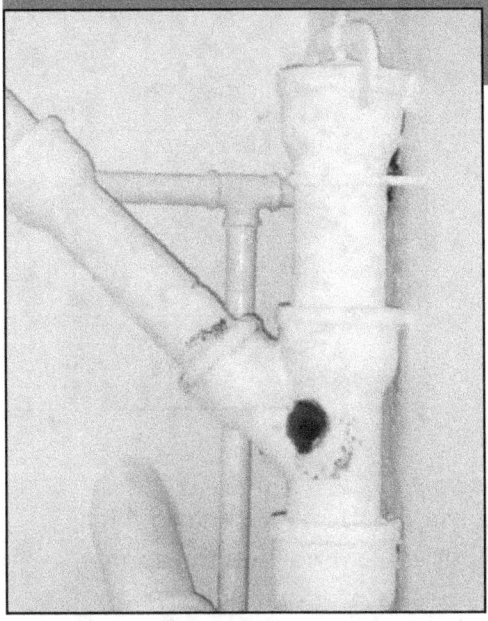

Rust is an indicator that condensation occurs on this drainpipe. The pipe should be insulated to prevent condensation.

Testing or sampling for mold

Is sampling for mold needed? **In most cases, if visible mold growth is present, sampling is unnecessary.** Since no EPA or other federal limits have been set for mold or mold spores, sampling cannot be used to check a building's compliance with federal mold standards. Surface sampling may be useful to determine if an area has been adequately cleaned or remediated. Sampling for mold should be conducted by professionals who have specific experience in designing mold sampling protocols, sampling methods, and interpreting results. Sample analysis should follow analytical methods recommended by the American Industrial Hygiene Association (AIHA), the American Conference of Governmental Industrial Hygienists (ACGIH), or other professional organizations.

HIDDEN **MOLD**

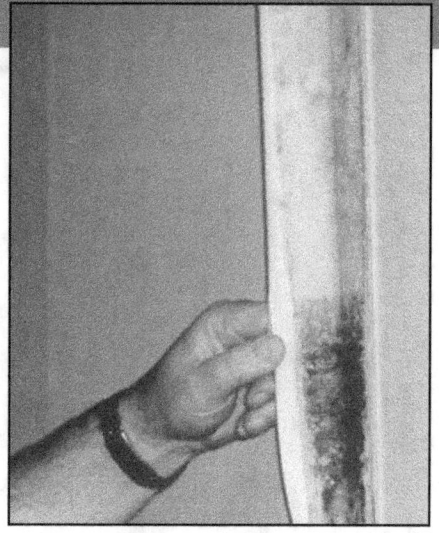

Mold growing on the back side of wallpaper.

Suspicion of hidden mold You may suspect hidden mold if a building smells moldy, but you cannot see the source, or if you know there has been water damage and residents are reporting health problems. Mold may be hidden in places such as the back side of dry wall, wallpaper, or paneling, the top side of ceiling tiles, the underside of carpets and pads, etc. Other possible locations of hidden mold include areas inside walls around pipes (with leaking or condensing pipes), the surface of walls behind furniture (where condensation forms), inside ductwork, and in roof materials above ceiling tiles (due to roof leaks or insufficient insulation).

Investigating hidden mold problems Investigating hidden mold problems may be difficult and will require caution when the investigation involves disturbing potential sites of mold growth. For example, removal of wallpaper can lead to a massive release of spores if there is mold growing on the underside of the paper. If you believe that you may have a hidden mold problem, consider hiring an experienced professional.

Cleanup and Biocides Biocides are substances that can destroy living organisms. The use of a chemical or biocide that kills organisms such as mold (chlorine bleach, for example) is not recommended as a routine practice during mold cleanup. There may be instances, however, when professional judgment may indicate its use (for example, when immune-compromised individuals are present). In most cases, it is not possible or desirable to sterilize an area; a background level of mold spores will remain - these spores will not grow if the moisture problem has been resolved. If you choose to use disinfectants or biocides, always ventilate the area and exhaust the air to the outdoors. Never mix chlorine bleach solution with other cleaning solutions or detergents that contain ammonia because toxic fumes could be produced.

Please note: Dead mold may still cause allergic reactions in some people, so it is not enough to simply kill the mold, it must also be removed.

Water stain on a basement wall — locate and fix the source of the water promptly.

ADDITIONAL **RESOURCES**

For more information on mold related issues including mold cleanup and moisture control/condensation/ humidity issues, visit:

www.epa.gov/mold

Mold growing on fallen leaves.

This document is available on the Environmental Protection Agency, Indoor Environments Division website at: www.epa.gov/mold

NOTES

NOTES

Acknowledgements

EPA would like to thank Paul Ellringer, PE, CIH, for providing the photo on page 14.

Please note that this document presents recommendations. EPA does not regulate mold or mold spores in indoor air.